Ek TEL in Afrikaans met VORMS

nommers 1 - 10

deur Nicolene Luff

Copyright © 2023 by Nicolene Luff

All rights reserved.

Paperback ISBN Number: 979 - 8 - 218 -18845 - 0

Copyright © 2023 by Nicolene Luff

All rights reserved.

No part of this publication may be reproduced, distributed, or transmitted in any form or by any means, including photocopying, recording, or other electronic or mechanical methods, without the prior written permission of the publisher, except as permitted by U.S. copyright law. For permission requests, contact Nicolene Luff at nicoleneinafrikaans@gmail.com.

The story, all names, characters, and incidents portrayed in this production are fictitious. No identification with actual persons (living or deceased), places, buildings, and products is intended or should be inferred.

This publication is designed to provide accurate and authoritative information in regard to the subject matter covered. It is sold with the understanding that neither the author nor the publisher is engaged in rendering legal, investment, accounting or other professional services. While the publisher and author have used their best efforts in preparing this book, they make no representations or warranties with respect to the accuracy or completeness of the contents of this book and specifically disclaim any implied warranties of merchantability or fitness for a particular purpose. No warranty may be created or extended by sales representatives or written sales materials. The advice and strategies contained herein may not be suitable for your situation. You should consult with a professional when appropriate. Neither the publisher nor the author shall be liable for any loss of profit or any other commercial damages, including but not limited to special, incidental, consequential, personal, or other damages.

Book Cover by Nicolene Luff

Illustrations by Nicolene Luff

2nd edition 2024

Paperback ISBN Number: 979 - 8 - 218 -18845 - 0

Laat ons taal voortleef deur jou!

driehoek

vierkant

sewehoek

agthoek

reghoek

ster

ovaal

pyl

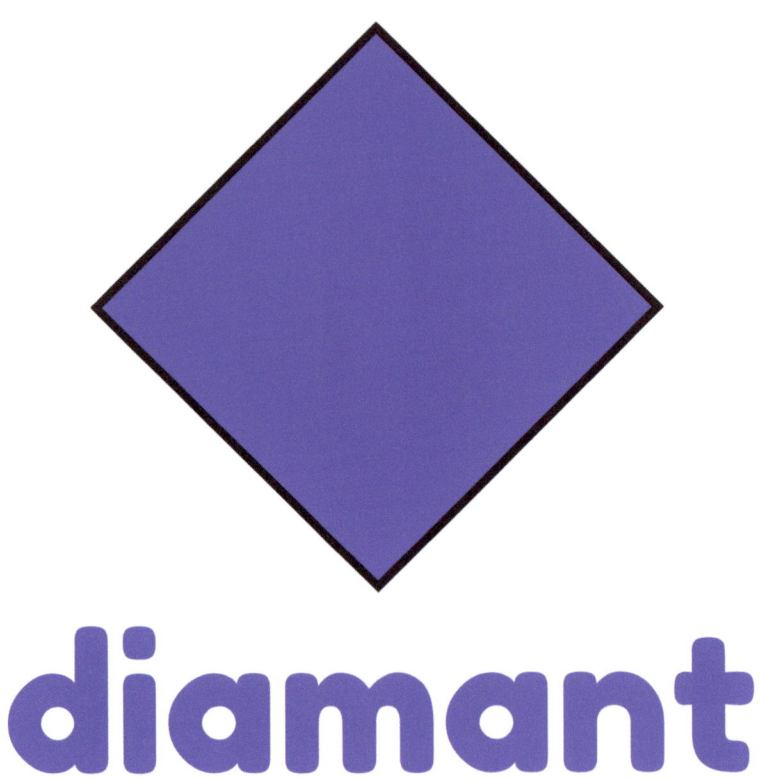
diamant

kruis

kubus

silinder

1
een

2
twee

3
drie

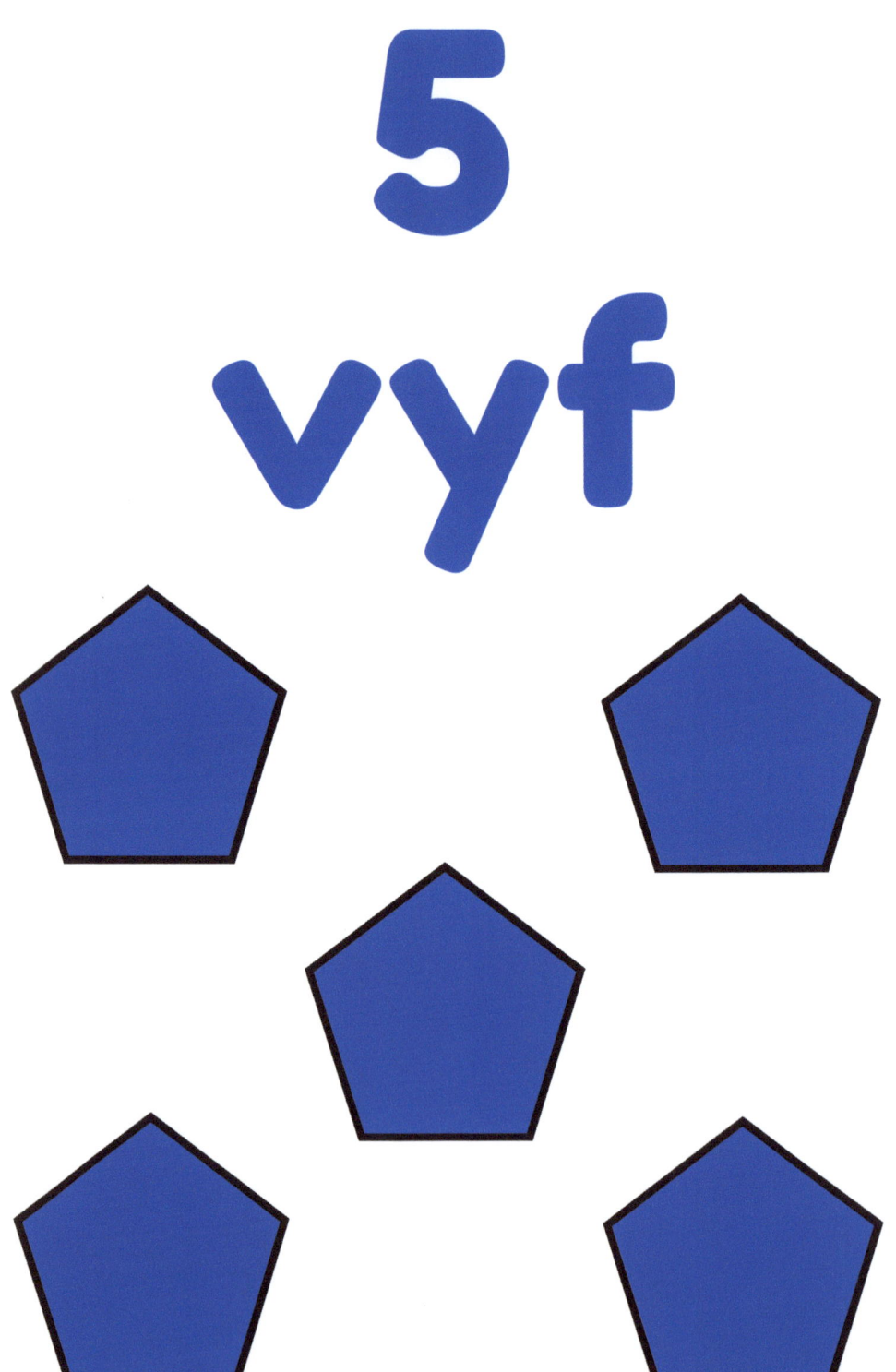

6
ses

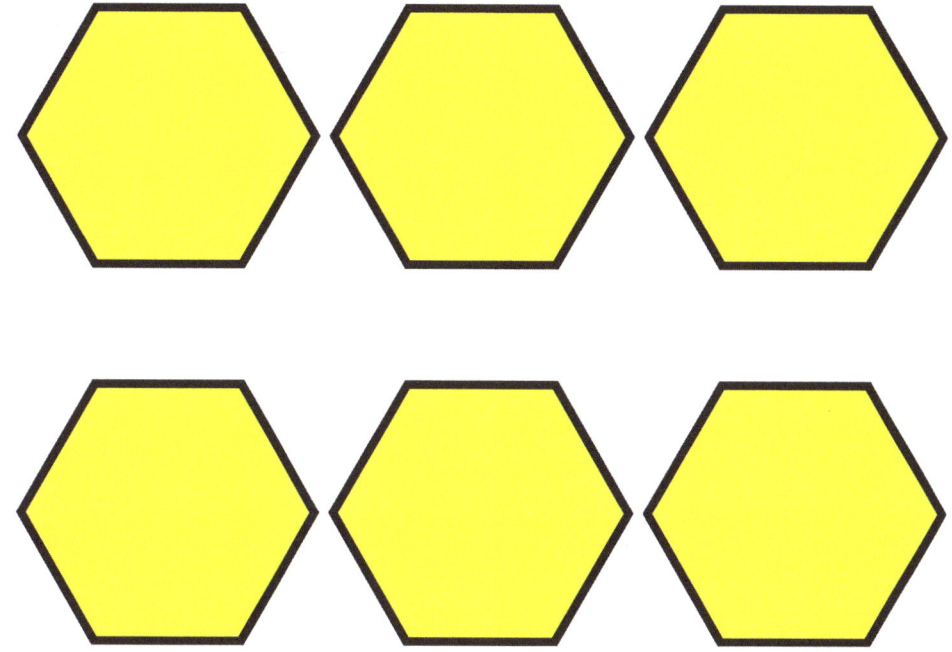

7

sewe

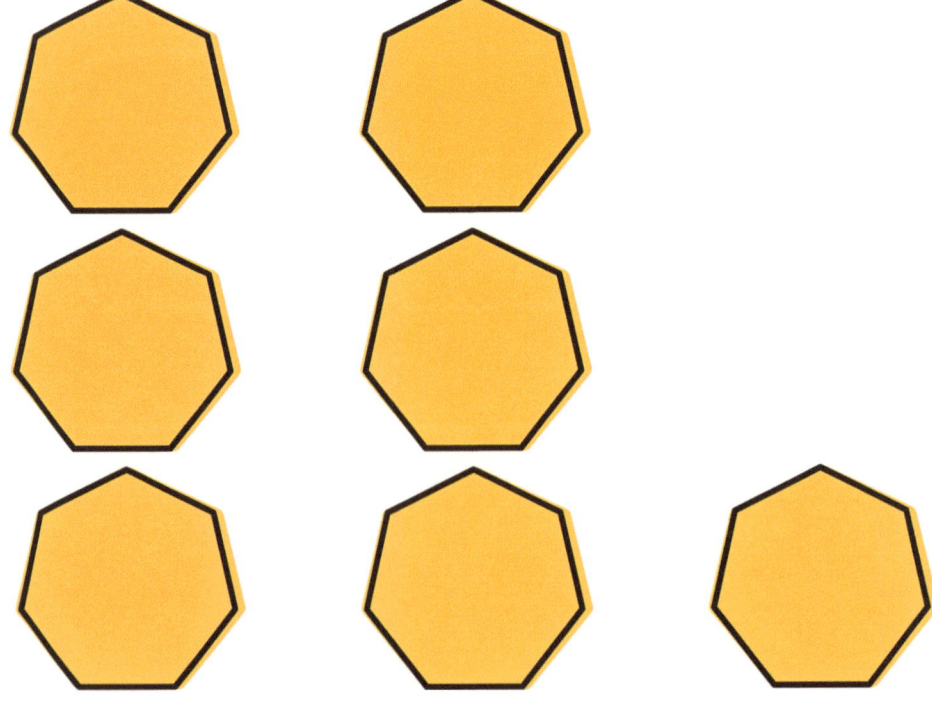

9
nege

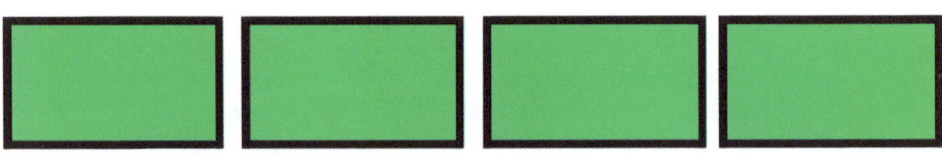

een
hart

drie
driehoeke

vier

vierkante

vyf
vyfhoeke

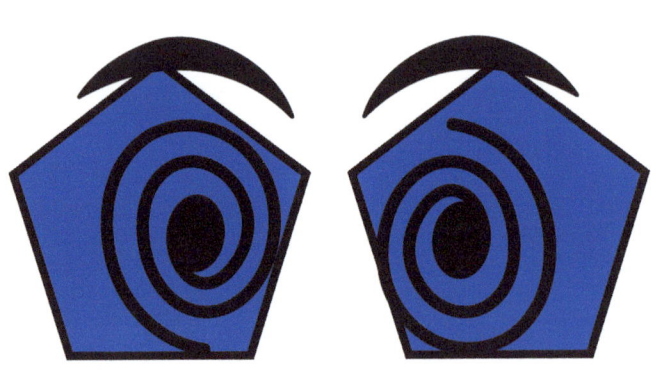

ses
seshoeke

sewe
sewehoeke

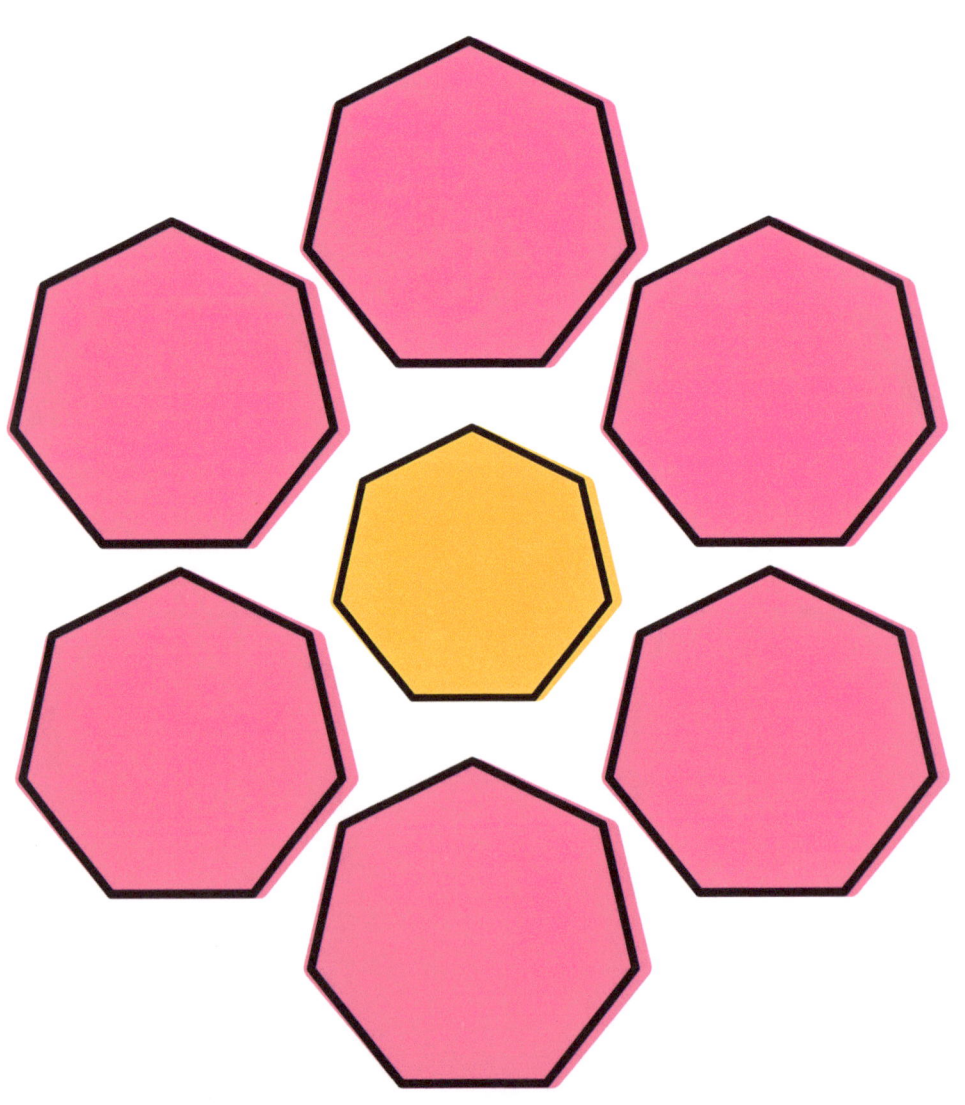

agt
agthoeke

nege reghoeke

tien

sterre

een
is minder as
twee

twee
is meer as
een

drie
is minder as
vier

vier

is meer as

drie

vyf
is gelyk aan
vyf

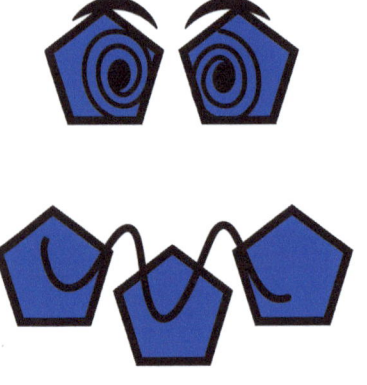
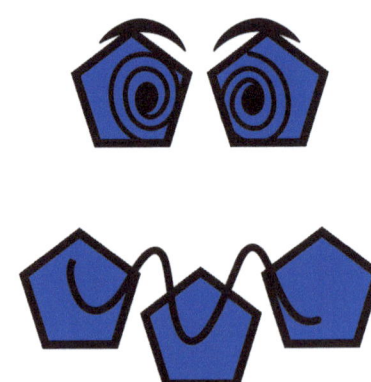

5 = 5

ses
is gelyk aan
ses

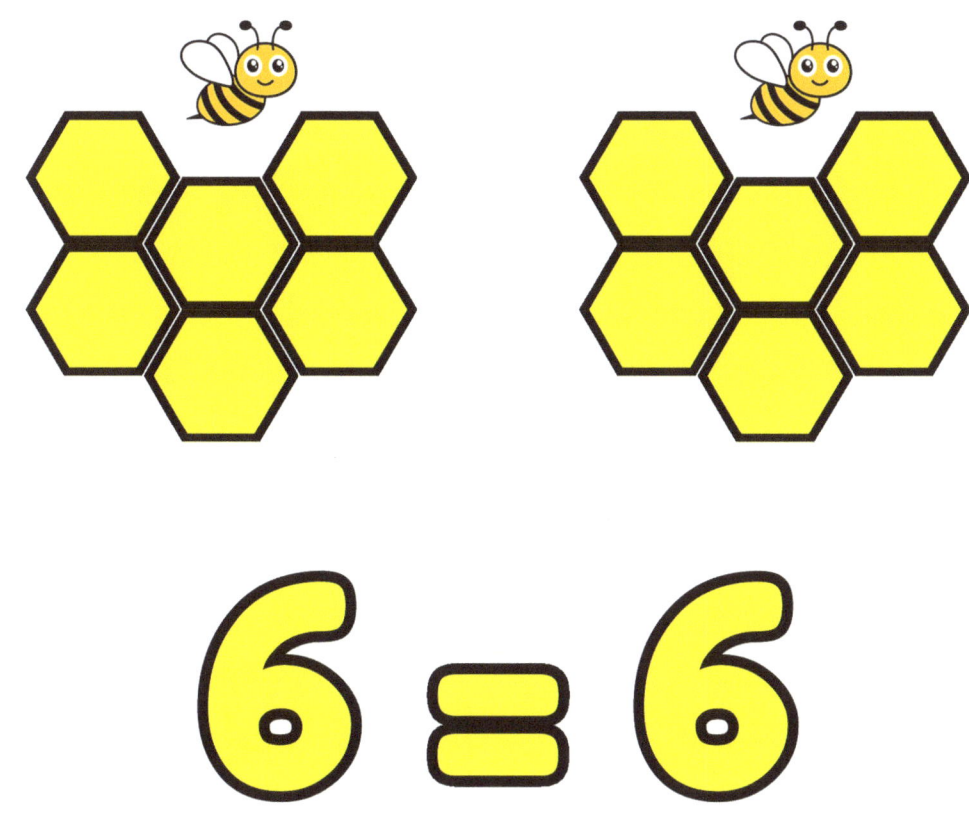

6 = 6

Die nommer sewe is <u>kleiner as</u> die nommer agt.

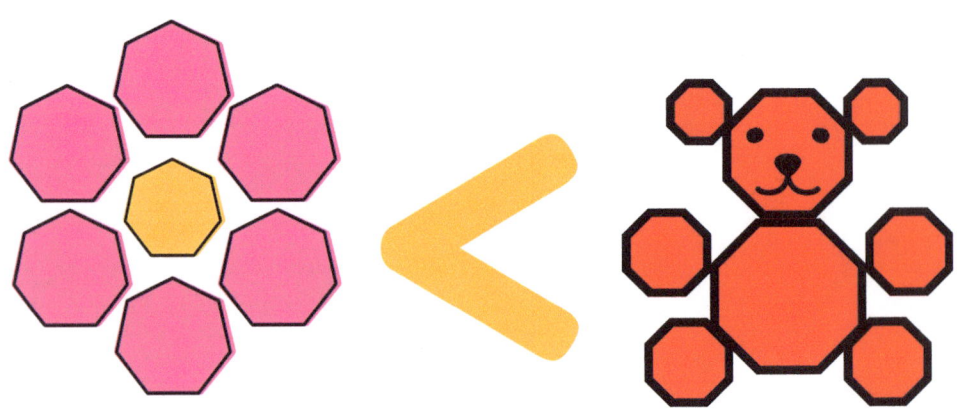

7 < 8

Die nommer agt is <u>groter as</u> die nommer sewe.

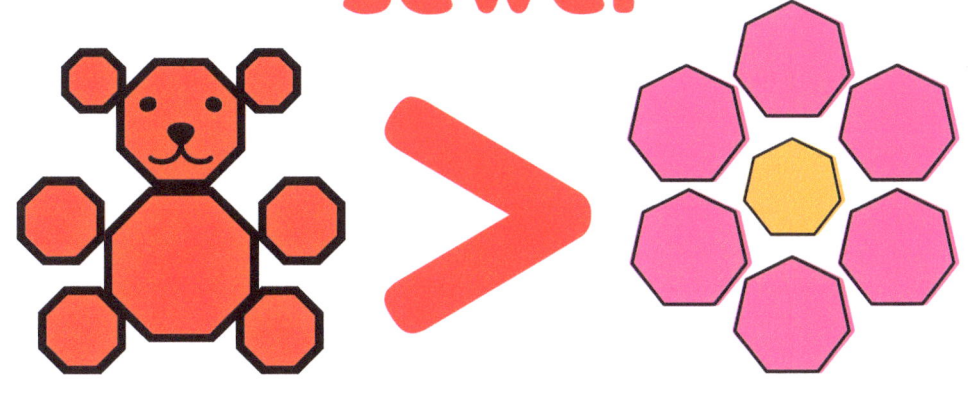

8 > 7

Die nommer nege is <u>kleiner as</u> die nommer tien.

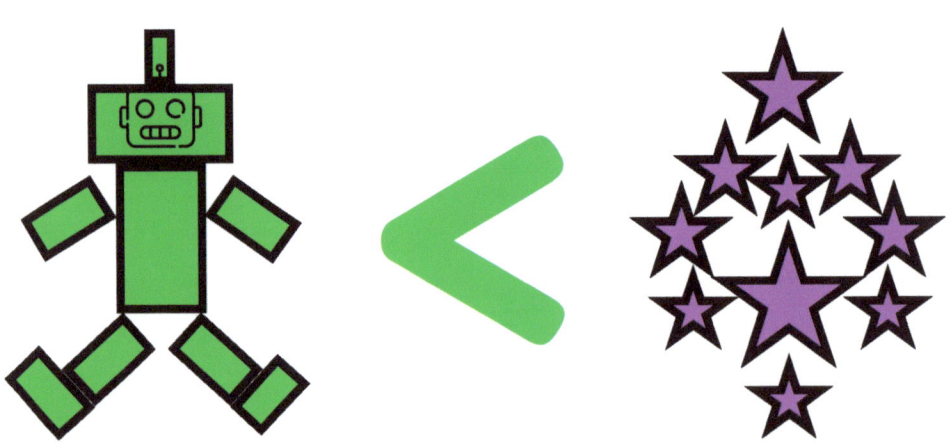

9 < 10

Die nommer tien is groter as die nommer nege.

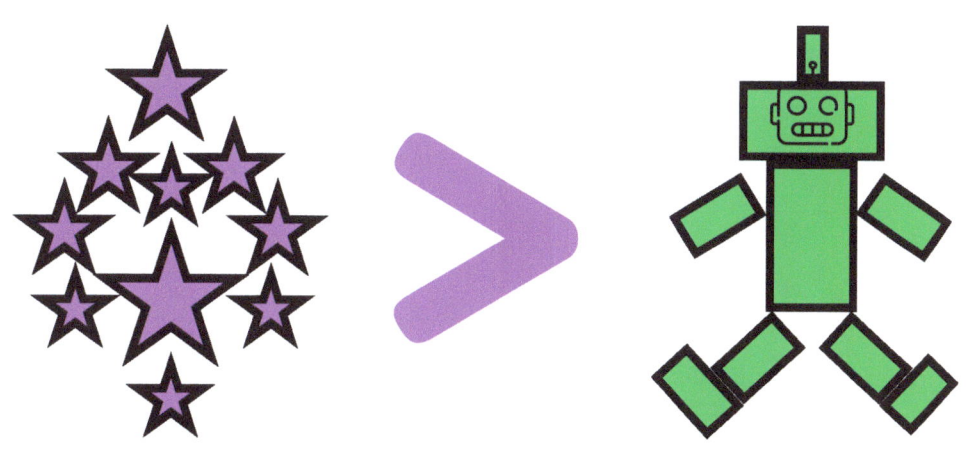

10 > 9

1 een

2 twee
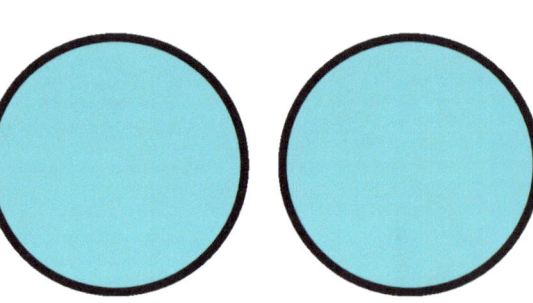

3 drie

4 vier
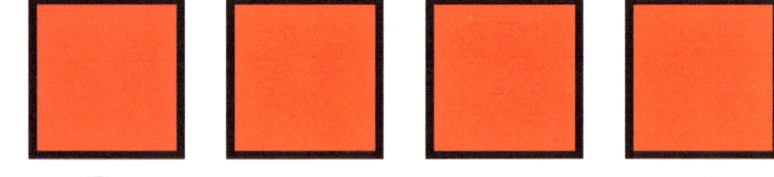

5 vyf

6 ses

7 sewe

8 agt

9 nege

10 tien

Daar is <u>een</u> beertjie met 'n <u>hartjie</u> op sy maag.

Die waatlemoen lyk soos 'n <u>driehoek</u> en het <u>vyf</u> sade.

Die stokkielekkers lyk soos <u>sirkels</u> en daar is <u>sewe</u> in die bottel.

Die boksie is <u>vierkantig</u> en daar is <u>nege</u> sjokolades daarin.

Die boekrak is <u>reghoekig</u> en daar is <u>tien</u> boeke op die rak.

This book is part of a series of books
written in Afrikaans.

Ek TEL in Afrikaans met VORMS, nommers 1-10
Ek TEL in Afrikaans met PLANTE, nommers 1-20
Ek TEL in Afrikaans met DIERE, wilde diere,
nommers 1-30
Ek TEL in Afrikaans met DIERE, mak diere,
tel met 10'e tot 100

& more!

Be on the lookout for other Afrikaans reading
& activity books!

Ek LEES in Afrikaans
Ek SKRYF in Afrikaans
Ek BID in Afrikaans
& more!

Find them on

Amazon
&
southafricantreasures.com

Follow us on Instagram
@southafrican_treasures

Thank you for your support!

www.ingramcontent.com/pod-product-compliance
Lightning Source LLC
Chambersburg PA
CBHW041307110426
42743CB00037B/30